ABOUT THE COVER
Artist: *Nele Azevedo*

To emphasize the crisis and importance of climate change and global warming in 2009, Brazilian artist *Nele Azevedo* carved thousands of these incredible tiny ice sculptures of men and placed them on the steps of *Gendermenmarkt Square* in Berlin, Germany.

Aside from being beautiful, *Azeveedo* created the miniature people to shed light on the effects of global warming and the *World Wildlife Fund's* warning that melting ice could cause water levels in our oceans and seas to rise.

The tiny men didn't stay sitting for long as they melted to visually demonstrate the point that *Azevedo* was attempting to make.

1

HUMAN POPULATION

and the case for

GLOBAL WARMING

Overpopulation and Destruction
of
Natural Resources and Ecological Balance
of the Planet Earth

by

P. Clay Sherrod
Arkansas Sky Observatories

Copyright August 2019

Revision 1

Arkansas Sky Observatories Publications
Available from: http://arksky.org/publications
Also available at Amazon, Barnes & Noble and other
booksellers
ISBN 978-0-359-80882-3

Introduction

It is somewhat oxymoronic that in the past several decades, for the first time in human history, the number of births *per capita* have decreased. This would seem like good news in terms of over population of the world we live on, but when you consider the exact "*capita*" on which the per capita is based, it appears that we have gained so little, if anything, in slowing the destruction of planet Earth's natural ecosystem that has provided all of the billions of species of biology – including that of mankind – with the abundance of both space and resources for what at one time must have seemed infinite.

Of all life on Earth, only humans can control their destiny; they have been given that ability, that right, because of the nature of our intellect which governs logic, processing of thought, and creativity. We are not the only creatures who can build a house – a bird, beaver, and insects can do that along with so many other living things. But we can go beyond the lodging for survival – we can stretch our creativity unlike other species to change the environment so that it suits our needs; if it is too warm we create air conditioning, if we do not like being rained on, we invent the umbrella.

It is that ability that sets humans apart from the rest of the living world. Further to that, the human specie, *homo sapiens*, is part of the animal kingdom and enjoys the ability to move and not be planted, like most of the biologically active portion of the plant kingdom. Nonetheless, we all are living, and we are all creatures of a planet....a very small planet and certainly one that is NOT infinite.

Now that we reach the 21st century we are experiencing just how finite our globe it. No longer is it flat and without boundaries; it is a sphere that has accumulated its resources and a living habitat for the evolution of biology over the past 3 billion years. The overall cosmic design of Earth was as near a perfect and ordering mechanism as could ever be imagined.

All species of the planet take from the Earth as they live the spans of their lives; the food, water, gases as resources as necessary; but the natural recycling ability of our planet from a geological, meteorological and biological viewpoint was nearly perfect in such a way that through the death of species which consume, their bodies also provided back into the ecosystem the resources needed for the cycle to continue.

Only with the introduction of the most intellectually capable living thing that the world had ever witnessed some three million years ago, did we see an entity on Earth that did NOT give back equal to what it consumed. For millions of years *homo sapiens* co-existed equally among the life forms from the standpoint of non-destruction.

It was when the ability and creativity evolved to a level where mankind became a selfish creature that the remarkable recycling ecosystem of this planet ceased to function like a perpetual motion machine. Exponentially resources broke down because a return to the earth was not there; and then followed

the explosion of population of these homo sapiens -
the Fission Population – which may ultimately lead
to self-imposed extinction.

Part I – Fission Human Population

You watch the ice of the Arctic circle over years....waiting. But you see no enemy ships, no threatening torpedoes, no foreign intruders.

There IS something, however, you do notice; a potential devastating threat for humanity that seems much less threatening and more sublime than machines of war.

Ice and the Survival of the Human Race

Yale University professor *Peter Wadhams* ("A Farewell to Ice", 2016) was an oceanographic scientist studying the thickness of the north polar ice sheet while aboard a British nuclear submarine in the early 1970s. Each year that he joined the crew aboard the vessel, he was noticing something peculiar and in his own words, "*....a stunning decline in the sea ice covering the northern polar regions. A more than 50 percent drop in the extent in summer and even steeper reduction in ice volume.*"

Just a few decades ago, ice 10 to 12 feet thick covered the North Pole, with sub-surface ice ridges in some parts of the Arctic extending down to 150 feet. Now, that ice is long gone, while the total volume of Arctic sea ice in late summer has declined, according to two estimates, by 75 percent in half a century.

In 2015, Wadhams recorded that the ice shelf over the Arctic Ocean was nearly the lowest ever on record, and he speculates that there is coming an "ice-free" North Pole during warm months, this possibly beginning as early as 2020. If it continues, he believes that there will be four months of no ice within the Arctic Circle within the decade and rapidly expanding to five months.

Having no ice covering the Arctic Ocean seems nearly impossible for most of us, and few have any idea the implications of just how serious this can be.

"As ocean and air temperatures in the Arctic rise, this adds more water vapor to the atmosphere, since warmer air holds more moisture. Water vapor is itself a greenhouse gas, trapping outgoing long-wave radiation and holding heat closer to the surface of the earth. With air temperatures rising in many parts of the Arctic by several degrees F in recent decades, water vapor concentration has gone up by more than 20 percent, adding to Arctic warming," explains Wadhams.

Both polar caps have been dubbed "The Air Conditioning of Mother Earth", these reflectively mirror-like ice caps helping to regulate the entire climate of Earth, and stabilize predictable – and livable – climate for periods of thousands of years.

In fact, it is noted that the once pearly-white brilliant reflective ice cap at the North Pole is now turning blue….not from contaminants, but from the sheer lack of any depth whatsoever to the sheet of ice that once covered the vast waters of the Arctic Circle.

The declining glacial ice of Greenland is now the largest single contributor to global sea level rise, its melting ice cap adding more that 72 cubic miles of water to the ocean each year. This would mean a sea level rise by the end of the 21st century of two to three feet minimum according to the Intergovernmental Panel on Climate Change (IPCC) .

But even this grim forecast is being revised upwards, with serious implications for the survival of low-lying cities like Miami, New Orleans, London, Venice, and Shanghai, or of potentially below sea level coastlines like Bangladesh. These major cities, and many others are very vulnerable to the rising of the ocean levels; consider the populations (2017 or 2018 whichever data is most current):

City - Population

Miami – 470,000
New Orleans – 395,000
London – 8,630,000
Venice – 262,000
Shanghai – 24,100,000
Bangladesh – 164,000,000

….or roughly about 200 million people now living in what could be under water by 2050.

Figure 1 – Global Temperature 1850-2017 / Courtesy Berkeley (CA) Earth

*　　　*　　　*

Noting the map in the following Figure 2, the ice coverage in the Arctic hit record lows through the winter of 2017. In March, when the sea ice should reach its largest area extent of each year, for 2017 it was lower than it ever had been in the history of satellite recording, over 40 years. This ice typically begins to melt in spring and by 2017 the melt began a month earlier than normal The Bering and Chukchi Seas along Alaska's coast remained ice-free longer into the fall than ever before recorded.

By November, 2017, NASA reported that two to four times as many coastal glaciers around Greenland are at risk of accelerated melting. Glacial coverage in Greenland is melting an average of **260 billion tons** of ice each year into the seas nearby, this of course resulting in the rise of sea levels world-wide.

The opposite pole is of course affected too, but is often overlooked because of its massive coverage and impressive sunlight reflective ability. The Antarctica cap realized the lowest ever drop in ice coverage in March, 2017. In July of that year – the middle of winter in the southern hemisphere - a **trillion-ton section** of Antarctica's Larsen-C ice shelf broke away from the south polar cap, and as with the northern losses, this attributes to rising of global sea levels – estimated to be up as much as three feet by 2050.

"There's no evidence that anything is recovering here," said *Mark Serreze*, the director of the NSIDC. "What we've seen historically is a downward trend in ice extent in all months. Superimposed on that are the ups and downs of natural variability. We're going to continue to head downward."

For the northern Arctic ice cap, the 2018 minimum was 1.63 million square kilometers (629,000 square miles) LESS ice coverage than throughout the 1981–2010 average ice minimum. NASA scientists *Claire Parkinson* and *Nick DiGirolamo* have calculated that Arctic ice cap has lost roughly 54,000 square kilometers (**21,000 square miles**) of ice for each year since the late 1970s – the equivalent to losing a massive portion of the North Polar Cap the size of Maryland and New Jersey.

Not just for one year….for every year since 1970.

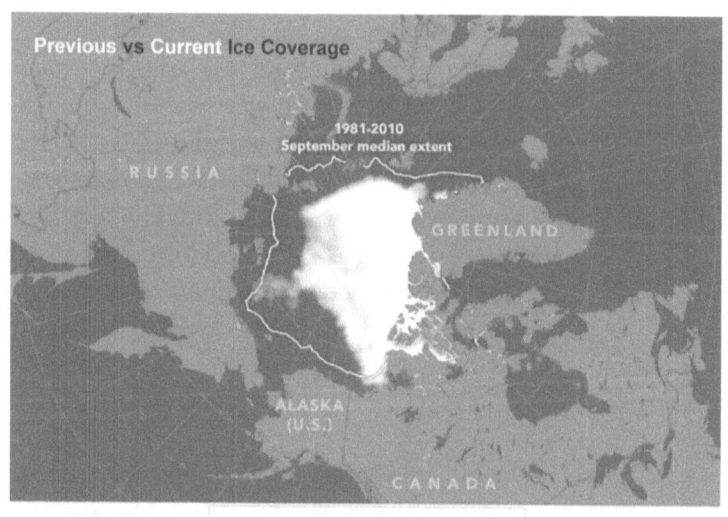

Graphic courtesy
earthobservatory.com / NASA

Figure 2 – Diminishing Ice Coverage in Arctic
Circle

GREENHOUSE GASES and the Effect From
the Biological World

There is an even larger problem in terms of the
effect on our world, one which could lead to
catastrophic changes in only the next few years.
This is the release of seabed methane — a potent
greenhouse gas — from the continental shelves of
the Arctic Ocean.

Yes, this is the same "methane" that the proponents
of the "*Green New Deal*" proposed by a few errant
free-thinkers in the United States government want
to curtail by eliminating all of the cows on Earth.
So called *Greenhouse Gases* are nothing new and
scientists have known of their effect since 1824
when Joseph Fourier showed mathematically that
the Earth's surface would be far colder if it had little
or no atmosphere; greenhouse gases such as Carbon
Dioxide, methane, water vapor, nitrous oxide and
ozone all contribute to the trapping of heat from our
sun and somewhat of an equalization of the Earth's
atmosphere over the course of seasons.

A BALANCED and steadily maintained proportion
of these gases create a natural greenhouse effect
which keeps the Earth's climate livable. Without it,
the Earth's surface would be an average of about 60
degrees Fahrenheit (33 degrees Celsius) cooler.
Levels of greenhouse gases have varied slightly
(sometimes dramatically for short periods) over the
Earth's history, but they had been fairly constant for
the past few thousand years.

This means that the average temperatures
throughout all of Earth have also stayed reasonably

consistent over that time. But that seems to have changed dramatically as the population of Earth has risen, particularly over the past 160 years.

The Human Factor

For the most part, planet Earth is content with the existence of BIOLOGY. For perhaps over a billion years, this world has sustained and created new remarkably efficient species of living things, from the basic virus to the organized manipulator: MAN.

Max Roser, in *World Population Growth* (Our World in Data, 2017) explains:
"200 years ago there were less than one billion humans living on earth. Today, according to UN calculations there are over 7 billion of us. Recent estimates suggest that today's population size is roughly equivalent to 6.9% of the total number of people ever born. This is the most conspicuous fact about world population growth: for thousands of years, the population grew only slowly but in recent centuries, it has jumped dramatically. Between 1900 and 2000, the increase in world population was three times greater than during the entire previous history of humanity—an increase from 1.5 to 6.1 billion in just 100 years."

Looking even earlier into the explosion of human population reveals that from 10,000 years before present (10,000 BCE) until the advent of the modern era (i.e., the time of the birth of Christ, 0 BCE) there were less than ONE MILLION people living throughout the entire area of the Earth. Not "billion" – million and far less than that in earlier times.

16

There is a hugely curious aspect to this seemingly out of control growth of the human race on our planet. While the enormous population continues to grow because of increased life expectancy and a substantially lower death rate per capita just in the past century (See Figure 10), the BIRTH RATE worldwide has dropped substantially, somewhat in part by family number restrictions in China alone.

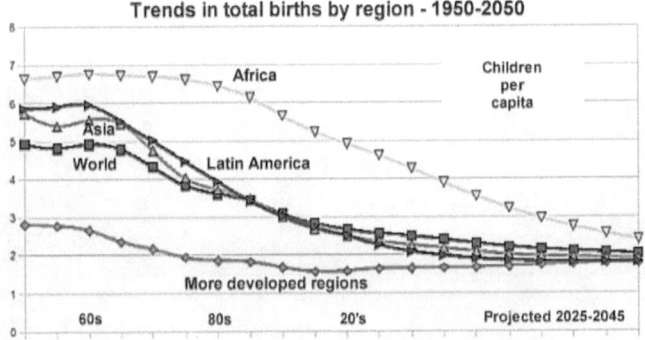

Figure 3 – Birthrate decline worldwide

Nonetheless, as seen in Figure 4, which reflects all aspects of birth, death and life expectancy, the population climb continues in an almost linear path.

By our year 2020, this number will have grown to over EIGHT BILLION – **eight thousand times** the number of humans starting at year 1 A.D (BCE). This suggests that we should look at the effects of industry, directly, as not as destructive to our global balance as the over-population of our planet.

This begs the question: *How many people can this planet support with its limited resources and limited surface area?*

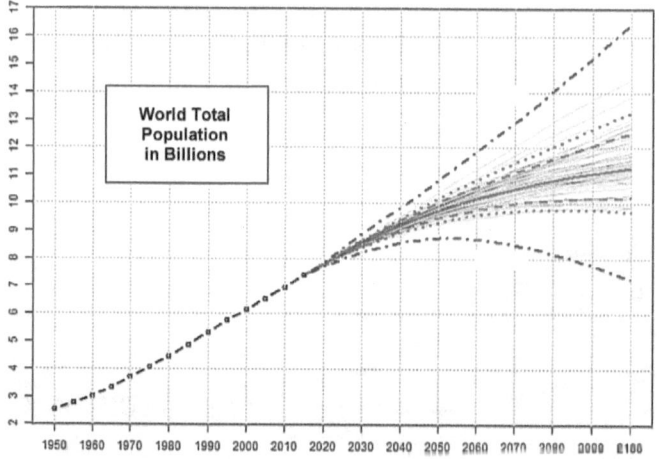

Figure 4 – Actual growth patterns and projections worldwide
Courtesy UN Dept. of Economic and Social Affairs 2017

A Finite Surface Area of Earth

So far as we understand scientifically, planet Earth is the only world within our space neighborhood that can support human life or life in any complex form. Of that world, only **197 million square miles** is <u>dry land</u>, suitable for habitation. That is only 29 percent of the total area of our planet, the rest being saline-bearing oceans and seas.

Of that 197 million square miles, nearly 33% - one third – is also unsuitable for habitation because of <u>desert conditions</u>, reducing habitable land to **131 million square miles.**

Reduce that space by 10% for all surface areas that are covered in <u>glacial ice</u>, and the habitable land suitable for human use declines to **111.3 million square miles**, still a seemingly large number….until we put all the facts together.

During the dawn of modern man, some 10,000 years ago, it would appear that there were some **111 square miles** for each human on the planet. That seems sufficient and then some.

Advance the civilization clock to today – around 2020 – and with 8 billion humans on the planet, we are left with a shocking **0.14 square mile** per person, or slightly LESS than eight acres, about the space needed to raise about seven cows.

Consider that New York City's Central Park is a whopping 843 acres by comparison.

But you might ask, if the density is so large, WHERE are all these people? The human race has partially solved this diminishing land crisis by living **vertically**, particularly in the largest of all cities such as New York.

These towering human storage facilities can hold a lot of people. Author *Andre Karam*, in <u>quora.com</u> writes:

"It depends on what's on that block [of the city]. If it's a high-rise apartment building, there might be 10,000 people or more. If it's an office building, there might be nobody "living" on the block, even though the office building might have 10,000 occupants working there."

Karam continues his assessment: *"My particular block is in a residential part of Brooklyn and is filled with low-rise apartment buildings. My own building has only 6 apartments, two of which have only a single inhabitant. So my building has fewer than a dozen residents. The entire block has 32 buildings (I just counted), some of which are commercial on the first floor and residential on the next two. So I'm guessing that there are between 500–750 people living on my particular block. But that's just a guess. By comparison, my apartment on Roosevelt Island (13 floors) had 500 units and about 2000 residents....."*

Is "going up" going to solve the problem with over-population? Really?

Note that the growth in human population is something of a remarkable phenomenon. This growth was not and is not linear....it is exponential, like the fission explosion of a hydrogen bomb. I invite you to compare the population growth graphic below (Figure 2) to that of the global temperature rise and note the similar growth during the same period of time.

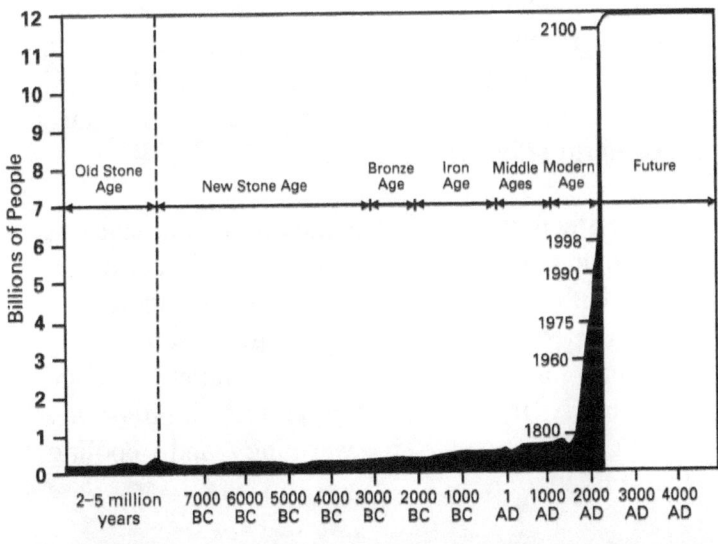

SOURCE: "Population: A Lively Introduction," Joseph A. McFall, Jr., *Population Bulletin*, Volume 46, Number 2, October, 1991, pages 1–43, Population Reference Bureau, Washington, D.C.

Figure 5 – Human Population Explosion Circa
1600 AD

*   *   *

It cannot be emphasized enough: the Earth is a finite planet with finite resources. It is incapable of "growing" simply because this biological entity known as "the human race" cannot keep its propagation in check. Eventually, perhaps quickly, this world will run out of room.

The Influence of Natural Factors

Human population and destruction of land area activity are not the only detrimental aspects that affect the Earth's climate. Volcanic eruptions add thousands of tons of ash and water vapor, as well as

21

harmful chemicals into the atmosphere and onto land every day; as the Earth has equalized since its beginning, this aspect has become less and less of a factor, but still is significant in terms of particularly short-term effects.

Cyclic fluctuations in solar radiation from sunspots and solar wind, can occur in very short periods of days, with cycles of 7, 11, and even hundreds of years in period. The Earth's position relative to the sun can also influence global climate change, particularly if solar activity is very high or low which can result in warming and cooling, respectively.

Even large-scale weather patterns such as El Niño and major meteorological changes that might be triggered from solar activity or flux in the Earth's magnetic field might result in fluctuations in temperatures worldwide for years or even decades.

It must be known that small warmings and coolings of the Earth's atmosphere when measured and averaged globally, will occur a single or one time events that can be tagged as such. The volcanoes will emit light-blocking ash that will temporarily hinder the passage of sunlight to the surface, thus resulting in a cooling effect for a short time, maybe as much as several years. Meteorologists are able to pinpoint the effects of El Nino to specific cycles, and even cycles of hundreds of thousands of years (i.e., the Ice Ages) can occur in repeating cycles that are NOT related to the steady climb in Earth's average temperature.

"Even though we're not setting records every year—and we don't expect to because of natural variability—we're not any where close to the averages we saw in the 1980s, 1990s and before," said *Walt Meier*, a senior researcher at the National Snow and Ice Data Center.

That being said, all climate models that are used by climatologists to monitor Earth's temperatures take those natural factors into account. When all of the "natural influences" are combined and considered, these changes in solar radiation as well as particulate matter, steam and noxious gases from volcanic eruption, have contributed only about *two percent* to the recent warming effect. Thus, ninety-eight percent of this rapid warming comes from greenhouse gases and other human factors such as overpopulation and abusive land use over time.

FINITE RESOURCES FOR INFINITE POPULATION GROWTH

Water, water everywhere....?

It may seem odd, but the fact is that there is the same amount of drinking water on earth as there has always been, only there are more people drinking it.

Figure 6 - Graphic Courtesy US Geological Survey
Clear drops represent ALL potable water on planet Earth

Freshwater makes up a very small fraction of all water on the planet. Only 0.007% of all water is drinkable. Most of Earth - nearly 70 percent – is covered by water, but only 2.5 percent of it is potable or fresh. The rest is salien/salt infused and ocean-based. Now take into consideration that just **1 percent** of our freshwater is accessible, the rest trapped in glaciers and snowfields. There is not much water available for 8 billion people.

It becomes somewhat distressing, or certainly should, to each of us when we learn of just how little water there is. As stated, the water is not "going anywhere" away from planet Earth, but there are 8 billion people using it instead of the less than 1 million just 2000 years ago.

So where is our water coming from? Certainly not the massive oceans the blanket our planet; that water is salenated - contains dissolved salts that cannot easily or inexpensively be separated from the liquid to make it potable.

Here is a breakdown of the only sources for potable drinking and consumable water that remain; you likely will be stunned by the fractional amounts available.

SOURCES OF POTABLE WATER

Fresh groundwater (may contain contaminants) – 2,530,000 cubic miles
Fresh water lakes – 22,000 cubic miles
Atmosphere (in form of humidity, clouds, suspension) – 3,100 cubic miles
Swamps of Earth – 2,750 cubic miles
Rivers – 510 cubic miles

Surprised how little there is? You should be and you need to be.

How much water does the average person on this planet use or consume per day? Each of us is using more and more water every day according to statistical studies, likely because of carelessness in the manner in which it is used. The average reveals that each person uses about **80-100 gallons** of water per day, the largest usage being that of flushing the toilet, followed by your daily shower or bath.

The inhabitants of the United States are the largest water consumer by far.

Worldwide, considering 8 billion people but eliminating an arbitrary 2 billion who may suffer from famine and drought, this will leave 6 billion humans each of which is using as much as 100 gallons of fresh water daily. Thus, our worldwide average daily consumption of water by all humans is approximately **600 billion** gallons.

One square mile of water contains 1,101,117,147,352 US gallons – about 1.1 trillion gallons. So, roughly estimating, we people of Earth use one square mile of water every TWO days.....about the amount of water that flows over Niagara Falls in a month.

And even knowing this, the average golf courses throughout the world are the top users of water – requiring between 100,000 and one million gallons annually for their daily watering, this according to the Alliance for Water Efficiency.

As we continue to allow the faucet to run while we text on our phones, or brush our teeth, you may want to remember that "water" is really a big deal. In fact: The Biggest.

Nothing on this planet will survive without fresh water, potable water. Many environmentalists and climatologists believe that fresh water will be more valued than gold or diamonds within this century. Perhaps wars will be fought over water rights; if conservation does not serve to provide the needs of a coming 16 billion people, then water rationing may have to serve that need.

I Think that I Shall Never See.....Another Tree?

Contrary to beliefs in your typical classrooms of higher learning, the world is slowly running out of trees. Deforestation is rapidly annihilating the forests that once dominated the landscapes of Earth. Whether for toilet paper or for the house you live in, our disregard for the forests of the world is slowly leading indirectly to Greenhouse gases out of control.

About 300 years ago there were on the order of as many as **10 trillion trees** throughout the world. Even then early culture consumed many of them without regard for replanting, such as the *Cahokians* of the Mississippi Valley did only 1000 years ago. Building great walls, palisades, structures and wooden monuments, those Native Americans soon found that it was no longer a mere walk to the woods to find great towering cedar or oak trees for their projects. Even farther fled all of the game animals that were the food source for their people, as the protective canopy of trees disappeared around them.

And nothing has changed since….except to perhaps get worse.

There are many educators and "experts" who believe they are in the know on the state of the world's forestation situation at this time; they are quick to tell you that replanting of tree farms (pines for rapid harvesting, not hardwoods) has led to more trees today than in any previous time.

They are wrong. Where once were those 10 trillion trees are now merely **THREE trillion** trees, according to global tree surveys done in 2016. Even more alarming is that we are using an astounding 15 billion trees each year for toilet paper, timber, clear cutting, housing and other human requirements.

In the United States alone, there are over 128 million houses and an additional millions of business structures. The number of trees required to build a typical 2600 square foot home is a minimum of 44; the homes now standing in the USA were once over **5 billion trees**.

According to the National Geographic Society, the world lost over **512,000 square miles** of tropical Rain Forest to human deforestation between 1990 and 2016. Most of this lost resource will remain lost forever. Perhaps someone needs to shake the person sitting next to him or her and explain a few very important facts about human survival as it relates to trees.

Let us start with the aesthetics of the trees as they relate to the solitude and sanity of mankind. Only people who have never entered a forest do not know the tranquility that overcomes the human spirit, whether in grief or happiness, when walking into a forested world. Home to thousands of species of life within every acre, there is peace and common ground between Mother Earth and humans within the woods of the planet on which we live.

Those same aesthetics are realized, but in a much different and more practical way, to all of the forest animals and plants that rely on the perfect balance of the woodland to all of the living structures – plants and animals – that live within the forest. It is habitat, protection, survival, food....indeed the forest is life itself for all that live within. Without the trees, there is no life.

Figure 7
Illegal deforestation of 3,050 square miles of the
Brazilian Tropical Rain Forest in 2018
Image courtesy BBC

In the tropical rainforests of South America, illegal deforestation is resulting in what likely will be demise of the most important ecosystem on the planet Earth. All animals on Earth depend on the vital gas OXYGEN for life....it is the breath of all that lives in the animal kingdom. The rainforest is responsible for over **20 percent** of all oxygen that is available on Earth....every molecule of it from the priceless trees.

Not just the rainforest, but all trees of Earth provide balance and life for pretty much all living things that have evolved on the planet. Without trees, there would not be enough oxygen for even a handful of humans to exist here. Each deciduous tree (broad leaf, like oak, maple, mahogany, teak, hickory) can provide over **260 pounds** of oxygen every year. Two of them can supply a *family of four for a year* with this breath of life.

But remember: we are losing 15 billion trees every year; those that we replant are not deciduous trees, but pines with such little photosynthesis capability as to not come close to the human requirements for oxygen on a yearly basis.

Note from the graph on the following page the correlation of population growth to deforestation. Notice the remarkable similarity to the earlier comparisons for global temperature increases and human population of Earth.

Not only do these trees give freely this vital oxygen, but they also assimilate Greenhouse gases that contribute to the ever-increasing warming of our planet.

In all of these factors leading to global warming and climate change, there is one dominant common denominator: uncontrolled   human   population growth.

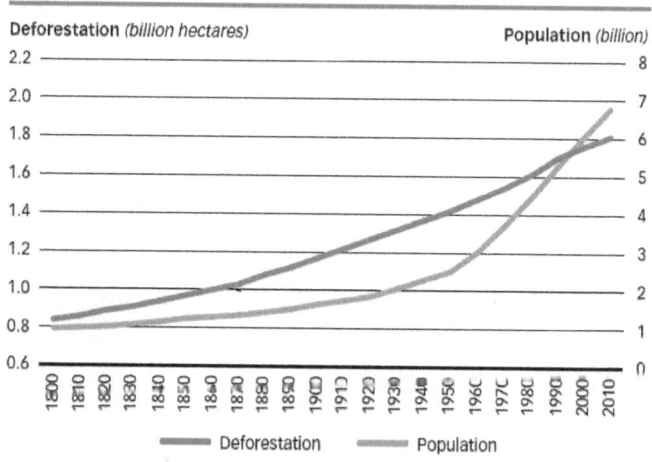

World population and cumulative deforestation, 1800 to 2010

Sources: Williams, 2002; FAO, 2010b; UN, 1999.

Figure 8

One such example of the severity of forest destruction is somewhat of a "catch-22"; a healthy green deciduous tree provides enough oxygen in one year for two people. While those two humans are inhaling and utilizing the oxygen that the tree provided them, they are required to EXHALE a virtual poison to the human body – carbon dioxide.

It just so happens that the tree once again steps in to save the human race; each human exhales about 2.3 pounds of carbon dioxide each day if activity is normal. That is a yearly total of nearly 840 pounds of carbon dioxide – a Greenhouse gas that is responsible for temperature increase of our atmosphere – for every human on Earth.

One healthy tree will utilize 48 pounds of the gas per year, not nearly enough to keep up with the exhaling population of 8 billion humans. For every human who wishes to breathe normally, it is going to require some 23 healthy deciduous trees to rid the atmosphere of his or her carbon dioxide.

Combine that effort with the additional release into the atmosphere of all of the human-related industries and activities from which either carbon dioxide or carbon monoxide are emitted, and the need for massive amounts of healthy forests is immediately realized.

While the present number of trees on earth are sufficiently providing oxygen for our survival, there are rapidly becoming insufficient resources in our rapidly diminishing forests to naturally eliminate the waste of both industry and mankind.

One final – but very important – point to illustrate about this quandary between the expansion of the human race and trees: *infrastructure resources.*

We have already illustrated that the world's population of human beings is spiraling out of control. In this early part of the 21st century it has once again become "popular" and trendy for western civilization families to encourage large families; two seems to no longer be enough children even though there appears to be such little family bonding in this modern age. We are in an era where four to six children are desired and even encouraged from one set of parents to another.

In other portions of the world, particularly the third-world countries and impoverished nations, there is little or no attempt made for birth control; efforts to provide contraceptives to uneducated populations has served little or no purpose, with condoms being used to transport water and worn as gloves.

Figure 9 – A massive heat collector

No human wants to give up his or her life and we do not want nor expect that to happen. Nonetheless, we have reached the critical mass of both resources and space.

Infrastructure Resources and Population

Consider this interesting paradox that nearly all humans are contributing toward daily.

Every human has the ability for logic – we are the only creature on planet Earth that can reason and rationalize. We can figure out ways to survive, we can create works of art, design great buildings and bridges, expand our frontiers simply through ingenuity. Nothing else living can do that, only us. A raccoon cannot, although he can figure out how to pop the lid on any garbage can. The rest of the living world lack the human ability to conceive concepts and develop ideas into reality.

Also, while on the subject of the raccoon we should note that the raccoon is not destructive to the environment, whereas the human is. Nor are birds destructive, snakes, bugs, spiders, elephants….nothing, but humans.

We are the only creature – and there are now 8 billion of us and growing – who can alter his environment to make it what he wants it to be. The only creature to intentionally create waste products and then carelessly throw them along a roadside. Over the countless years, all other biology actually contribute to the success of the world; they give back, even with the deaths of their bodies which decay into soil for future living things. Humans do not even do that: we cremate or embalm our dead bodies so that they are worthless flesh that enriches nothing of the world.

Even in death, the human is a selfish creature.

So let us examine the "human drive" and link it to the chain of Greenhouse gases and the rapid deterioration of our planet that is leading to the melting of the polar ice caps. Following is an exercise to understand the human desire to "keep up with the Jones'" and this drive appears implanted in all of us, myself included.

Babies are born and rapidly grow to maturity. Along the way, all traits of human rationalization begin to develop and the influence of society and peers begins to take an equally dominant role in thought processes.

Every one of the children grow up and if possible one of the first things we all want is an automobile or means of transport, usually requiring polluting fossil fuel propulsion. Soon, the human finds more mature needs, perhaps a home for a growing family, and seeks a place to put both the automobile and the home. To make place among the many other homes of other humans require "neighborhoods" and "cities" and these developments require that growing living green things – many times deciduous trees – be cut down to make room for each domicile in the neighborhood.

So we at some point, let us say, have a family of four now living in a small parcel of land where trees once stood producing oxygen and purging our air of poisonous Greenhouse gases.

The people within the household, as well as all neighbors, of course want a place to park the vehicles that they own when they are not out burning fossil fuels in them and emitting noxious Greenhouse gases. So green grass and plants (also capable of photosynthesis to produce oxygen and reduce poisons) are scraped away and cement is poured in its place.

But that is not enough: there must be connections throughout the land areas where the vehicles can rapidly traverse from one destination to another – highways and streets, all requiring the removal of green growing things and replacing that with carbon dioxide and Greenhouse gases as well.

Shops must be built, schools constructed, stadiums erected, bridges for crossing those pesky water sources....more concrete, less trees. And, yes, we must have huge parking lots made of asphalt or cement on which to park our automobiles at these shops, schools and more.

Aside from the marvelous process of photosynthesis, green plants have a remarkable capability to cool down the immediate environment in which they reside. Things are simply cooler in the woods than they are in a parking lot. *But stop*! We are reaching the point where there are more parking lots, houses, roads, buildings and bridges than there are forests.

I invite you to drive through any large metropolitan area on an Interstate Highway in the middle of the night on your next trip from one destination to the other. Watch your automobile thermometer – most are quite accurate. As you approach the large city, take a look at the night time temperature; perhaps it may read "71 degrees"; but keep watching. As you pass through the most dense part of any metropolitan area, that temperature will rise usually at least THREE degrees, all of that being heat which was retained in the cement and steel during the hottest part of the day.

While we are on the subject of roads and their influence on the diminishing forests and the ever-increasing global temperatures, join me in a quick look at the resources that the earth provided to construct the Interstate Highway system. These are resources for which trees were bulldozed to dig the earth, resources which required millions of years for the Earth to compose, and resources that we will never replenish in the future of mankind.

As of the first decade of the 21st century, there were 97,000 miles of Interstate Highway traversing just the United States. Those highways, made of cement on treeless avenues hold over 55,000 bridges of cement and steel. The stone used simply for the roadways is enough to build 700 more pyramids in Egypt. The cement poured on top of that roadway is enough to have constructed SIX sidewalks to the moon – 1.4 million miles.

And every inch of this required the removal of an Earth-saving green growing thing...but we can get from that one destination to another very quickly.

You see, the human specie is perhaps the most selfish of all in all respects other than survival. Yet humans have the capability through logic and creativity to ALTER the environment around them to survive, or at least survive more easily.

One more example that demonstrates the importance of infrastructure over the future of the climate of the Earth, that being San Francisco's Golden Gate Bridge: A marvelous feat of engineering and design, but one wonders if it really had to be built. Is there not a long way around getting to Oakland?

The Golden Gate Bridge is comprised of a remarkable 80,000 miles of steel girders, cable, rebar, bolts, nuts and other gear. Every ounce of that material was produced by Mother Earth long before mankind ever walked anywhere near the oceans. Even with the great bridge completed, worldwide we still continue to pillage the Earth of 43 million tons of iron ore annually, all from strip mines where once stood the oxygen-rich, poison-devouring great trees of Earth.

What becomes of this Earth if we continually drain it of its components and alter it in a way that only mankind can do?

Reaching the Point of No Return

Early in the 1970's I traveled throughout the United States and lectured on the dangers of uncontrolled population and the limited resources available on this finite planet Earth. Perhaps I was a Pioneer in that respect, but they always say that you can tell the "...true pioneers by the number of arrows sticking out of his back."

And that is the way it was.

In 1972 people simply did not give any mind to climate change; we had Vietnam, acid rock, hippies and the beginning of political posturing.

From then until today I emphasize that the root condition that leads to the changing of any world in the scale that we are seeing today is NOT a natural phenomenon like a solar cycle or posturing of the magnetic field of this planet. The root cause is very simple:
OVER-POPULATION.

The solution to the problem is two-fold:
1) fix the population problem
2) find a way to fix a broken planet.

Neither of which is at all simple and frankly neither of these parts of the solution is likely ever going to take place. It seems that we have stretched our finite Earth beyond her abilities to support and maintain our lifestyles. The human race has outlined, staged, and produced perhaps the death sentence of all biology.

My efforts in the 1970's and those of others are not unlike those of today; each generation of humans believes firmly that the issue does NOT involve them and that somehow this entire idea of the human race driving itself some what *purposely* to extinction is abstract and absurd.

Consider that the problem has become even more pronounced in the 50 years since the 1970's through the advent and progress of modern medicine. Quite simply humans are living longer; since just 1970 the life expectancy has risen from 67.1 years (male and female combined) to an incredible 82 years for both male and female by 2018 (source: Berkeley Medical).

This becomes even more of a factor when you learn that life expectancy in 1500 Europe was only 35 years of age.

Figure 12 – Life Expectancy in Years 1858-2010+

Obviously the human race is seeing a longer life span with fewer deaths. All the while, in relation to the total birth rate which has dropped significantly over the past century per capita, the total population still surges higher and higher with all aspects considered:

- total decline in birth rate per 1000 humans
- total increase in life span of average human
- significant decrease in annual death rate

The sad fact is: the decline in the birth rate at present per capita simply cannot offset the population surge and thus the overall occupancy of the planet continues to rise (Figure 4) and the planetary occupancy by the human race continues to grow at a linear rate.

If the human race suddenly today ceased just where we are, then biology would recover; it always does once left alone. But that is not likely to happen and the fantastically silly overtures of "New Green Deal" and "Going Green" is juvenile and ineffective at best. It may simply be too late.

The fact is: life existed on planet Earth long before the appearance of mankind and sustained just fine without altering the planet to do so. Unless the only creature which CAN detrimentally alter his environment is the human, if our specie disappears and leaves just any aspect of biology intact before doing so, biological species that do not have the ability to reason and logically create will once again thrive on the planet.....depending on what the human race leaves behind.

An Unpleasant Prognostication

The British scientist/submariner Peter Wadhams perhaps first provided illustrative proof of just how quickly climate change was altering the northern Arctic Circle. The rapid seasonal decline of the ice caps into the fall are telling, but the summers have really provided scientists with evidence that we may be losing the North Polar Cap ice, and this loss may be permanent. As early as 2030 the Arctic Ocean could lose essentially all of its ice during the warmest months of the year—a radical transformation that would destroy virtually all of the Arctic ecosystems and disrupt or destroy many northern communities, if not all communities along the coastal areas of Earth.

Throughout its history, we know that the Earth has worked though cyclical climatic changes, some of which happen at fair regularity. The last of these was our last **Ice Age** when the difference between average global temperatures today and during those ice ages was only about 9 degrees Fahrenheit; the swings have tended to happen slowly, over hundreds of thousands of years. We are aware of these and we know that they have absolutely nothing to do with anything that the human race might be doing on our planet.

But now concentrations of Greenhouse gases are rising dramatically – because of mankind's industry as well as human overpopulation leading to the destruction of the cycle of photosynthesis. Earth's remaining ice sheets over Greenland and Antarctica are shedding their glaciers at an alarming rate that increases with every passing day. That extra water could raise sea levels significantly, and quickly. By 2050, these melting glaciers will provide enough volume in the oceans and seas to raise the sea level by as much as 2 or more feet.

The cycle does not manifest itself just in melting ice caps and homeless polar bears. The rising temperatures will ultimate lead to *very extreme weather*, something that meteorologists are already seeing yearly in short term patterns: volatile storms instead of gentle rainfalls, increases in cyclonic activity from rapid upthrusts of heated air in the form of tornadoes and hurricanes, unpredicted droughts where none should be.

Indeed, the entire climate of planet Earth may shift creating deserts where none were before, swamps amidst the sands of Death Valley. With such meteorological changes come obvious side effects in the form of shorter growing seasons for crops for our 8 billion people, limited grazing and habitat areas where wildlife can thrive and propagate.

And ultimately, we are facing the possibility of the loss of precious water, the one substance that is required by all of biology to survive. The Earth can and will survive without potable water; the world of biology cannot. Just from the melting of the polar glaciers into fresh water, we have lost millions of cubic feet of water already as the pure essence falls diluted into a salty sea.

Has the cycle reached its "critical mass" and now unable to be reversed? Will popular social efforts such as "Going Green" help in any way whatsoever at this point in a global evolutionary crisis? In only a few – perhaps two – generations of the human race might we know the answers to whether the human race will have a planet capable of sustaining life without ever leaving this world.

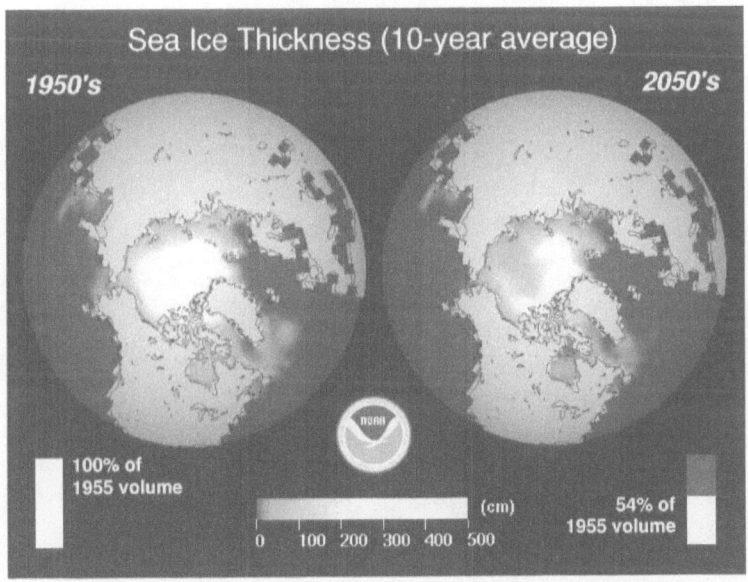

Figure 13
North Polar Sea Ice Thickness Decline –Century
Courtesy NOAA

Part II
Climate Change & Global Warming

Overview

Massive country-sized glaciers fall magnificently into the polar waters around them, one by one, eroding the great masses of ice that we have come to know as "polar caps" of Earth. Massive shelves of prehistoric in frozen state are finally giving way to the warming blanket of air surrounding our planet, the results of which are going to be catastrophic for future generations of all living things. The cause of the glacial erosion is very simple: human population explosion and the results of human demand and consumption.

The way to stop this: non-existent. The nonsensical "going green" movement should have begun in the early 1960's when scientists first sounded the alarms about this potential global disaster; putting into place an even more ludicrous plan such as "The Green New Deal" is nothing but political rhetoric…it is not a solution, nor any steps toward a solution.

In fact, the problem is so large – planet sized, so to speak – that no effort short of the elimination of the human specie will stop the cycle.

Global Warming

As blizzards blanket northern locations, some the worse on record, each year, it is quite easy for the majority of the lay public to confidently assess that there is no such thing as "global warming" or put

another way: Climate Change. Looking at any specific location covered in record breaking ice for any short time is nothing less that watching a pesky fly at the tip of your nose while the charging hippopotamus has targeted you for annihilation.
The fly means nothing; the hippo everything and its momentum cannot be stopped.

According to NASA's Goddard Space Flight Center, the average temperature of the entire globe of Earth has steadily risen by 1.4 degree F. since 1880; at first the rise was gradual and only slightly less than a tilt from linear; but looking at Figure 1 (Berkeley Earth www.berkeleyearth.org) the sudden uptick in annual warming for both land and water is obvious beginning about 1975-1980.

Figure 1 – Courtesy www.BerkeleyEarth.org

Often overlooked in climate discussions by media outlets and casual conversations about the implications of the warming trend is the even greater long-term influence of **ocean warming** which has become increasingly significant and dangerous for the future of mankind.

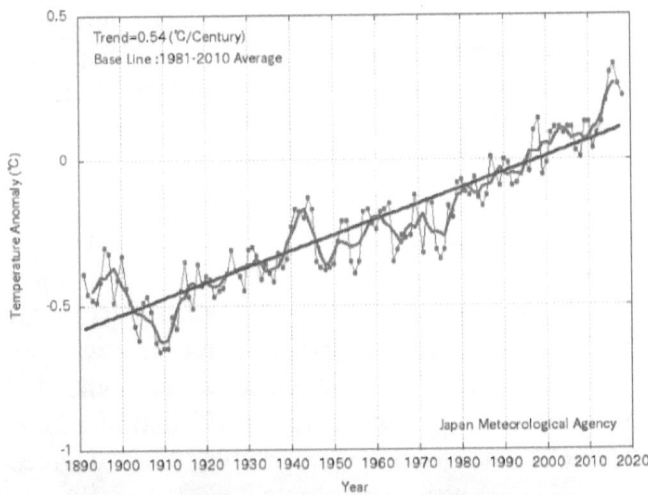

Figure 2 – Ocean Temperature Increase, 100 years
Courtesy Japan Meteorological Agency

As can be seen in Figure 2, the one hundred year trend of ocean/sea temperatures is also on the rise, gaining about one degree C. over the past century; this increase is nearly linear and constant at this time. Nonetheless, such a warming of the waters will have a catastrophic long term effect on a chain of events that cannot be reversed by simply "going green" as I shall discuss.

Warming water exponentially affects the overall climate in several ways:

1) the evaporative index of the oceans and seas in terms of total surface area of water evaporating into the atmosphere increases;

2) water, being denser than air, retains heat far longer and this warmth circulates with the

circulating currents of the oceans, including into the polar waters of the planet;

3) increased warm waters have a natural ability to radiate and dissipate heat into the surround air and land areas, thus warming glacial regions to what may be an alarming degree.

In fact, this warming near the arctic waters is dramatic and should have served as a wake-up call to Earth as early as 1960, when the average water temperatures began rising exponentially and continue to do so. In the past six decades, the average temperature of the arctic ocean – vital for maintaining the northern ice shelf and northern glaciers – has risen a remarkable three degrees Centigrade. (Figure 3)

Figure 3 – Showing the dramatic temperature increase of the Arctic since 1960
Courtesy www.berkeleyearth.org

We see these facts and this arctic warming, but why should this be any cause for the future of the human race? No one lives in that region, right?

Glaciers and Ice Caps – Earth's Natural Reflectors

Figure 4, below, demonstrates graphically the tonnage decrease in the SOUTH polar cap from 2002 through 2017 – ice loss of over 127 gigatons each year over a 15 year period.

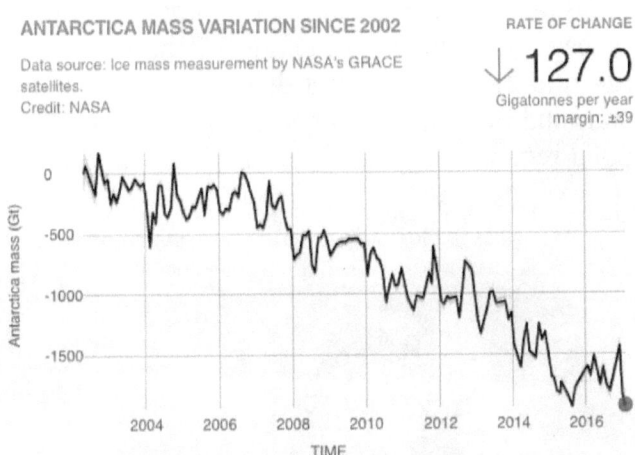

Figure 4 – Mass of Ice Lost from South Polar Cap in 15 Years
Courtesy NASA

The South Polar Cap – Antarctica – is an overwhelming large land of thick and reflective ice; in fact, it serves as somewhat of a thermostat for planet Earth in that it reflects about 12 percent of all sunlight the planet receives from the sun; imagine the increase in heat absorbed by our planet if this huge natural reflector did not exist.

The loss of frozen ice from this cap, melting fresh water into the salt of the oceans and seas, is depleting two things: 1) fresh water suitable for consumption; and 2) the reflective ability of our planet to maintain a constant balanced temperature.

The South Polar Cap is so huge that most people would not recognize the remarkable losses shown above in the 15-year period. The North Polar Cap – the Arctic – is far more obvious in terms of visually realizing our loss.

Since the late 1880's the decline of ice mass at the North Polar Cap is nothing less than startling. Measured as Ice Mass Concentration, Figure 5 reveals the dramatic shrinking of the northern ice into the 21st century, this due nearly entirely from warming of both the waters and air surrounding our polar regions.

Figure 5 – The dramatic shrinking of the North Polar Cap, 1879-2012
Graphic courtesy Woods Hole Research Center (WHRC)

As previously mentioned, the demise of ice at the Earth's poles has serious consequences for the future of the Earth, both in terms of climate and in habitability, among these:

1) ice melting increases the sea level to substantially higher levels that coastal population centers can tolerate;
2) pure fresh water in the form of ice melts into the non-potable salt waters of oceans and seas;
3) reflectivity of the Earth drops significantly, thus increasing the overall heat absorbed from the sun, rather than being deflected.

Sea Level Changes from Melting Ice

There is no doubt that there have always been fluctuations in the sea level across the globe, these changes due in part to geologic events for major global changes and many times meteorological events specific for local increases.

However, the sea level across Earth has been steadily rising since the beginning of the 20th century, to an average of roughly 6-8 inch since 1900. (*USGCRP (2017)*. *"Climate Science Special Report. Chapter 12: Sea Level Rise". science2017.globalchange.gov. Retrieved 2018-12-27).* In fact, projections for sea level increases up to **three feet** by the end of the 21st century appear to be firmly supported in the scientific data (*Global and Regional Sea Level Rise Scenarios for the United States (PDF) (Report) (NOAA Technical Report NOS CO-OPS 083 ed.)*

In 2014 the Intergovernmental Panel on Climate Change attributes 44% ofthe rising sea waters to the melting of glaciers worldwide, including those of Greenland and Antarctica. As the glaciers melt from frozen to liquid state, the fresh water contained in them mix into the salty oceans and seas, increasing sea levels and diminishing even further the amount of potable water on the planet.

Our oceans serve as a thermal valve against overheating, being responsible for absorbing and retaining over 90% of the excess heat that the Earth receives from the sun, and generates internally from geological processes.

The melting of the Antarctic ice sheet is particularly worrisome and dramatic, with over 4.3 gigatons of ice melting annually into the waters surrounding it. Equally obvious is the melting of ice over Greenland, attributing 15% of the total water added presently to the oceans around it; in fact it is now accepted (*Mosbergen, Dominique (2017). "Greenland's Coastal Ice Cups Have Melted Past The Point Of No Return". Huffington Post*) the Greenland glacial demise has reached the point of no return: the melting is irreversible and ultimate the Greenland glaciers will soon be gone.

Keep in mind that over ten percent of the world's population currently lives around the coastal areas of the world that are earmarked for coastal flooding from sea level increases during the next century; thousands of islands and island nations will be under water if the current projections for sea level increases hold.

Since the last ice age, some 20,000 thousand years before present, the sea levels across the globe have risen over 410 feet, this being primarily of course from the melting of the huge quantities of the Laurentide Ice Sheet and other temperate glaciers.

Nonetheless, the waters continue to rise from unnatural root causes, the principal being the influence of over-population and the carbon dioxide emissions from the "progress" of modern civilization.

Reduction of Fresh Water by Melting into Salty Oceans

There is a limited supply of fresh water available on this planet. Of all Earth's water, 96.5% of it is salt water and will never be suitable for drinking or human consumption. The remaining 2.5% is potable and found in only a few sources, among them glaciers. At present, nearly 69% of all fresh water is locked up in frozen form in the world's glaciers; additionally a small percent of fresh water is found at high altitude glaciers – mountaintops – which have also been influenced by climate change and are melting at a very rapid pace.

In the case of high elevation glaciers, most of this fresh water will melt into small creeks, then rivers and larger bodies of water and remain as potable water for long periods. However, the 69% of glacial water – when melted – flows into the oceans and ultimately will be mixed in with non-potable salt water.

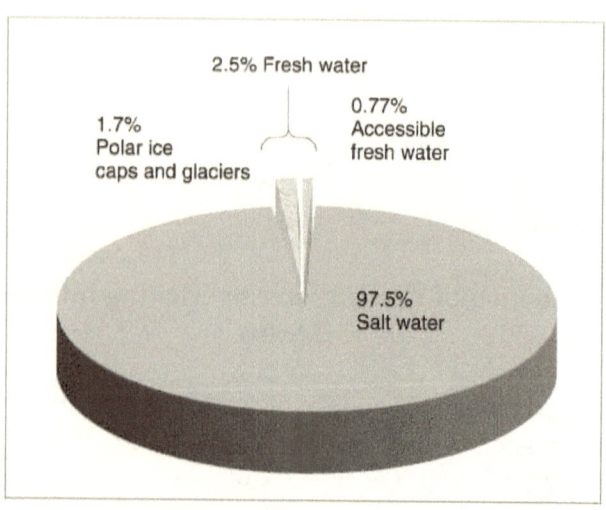

2.5% Fresh water

0.77%
Accessible
fresh water

1.7%
Polar ice
caps and glaciers

97.5%
Salt water

Figure 6 – Demonstrating the availability of fresh water

Lowering Earth's Reflectivity (reduction of albedo)

All objects – planets, moon, Earth – in the solar system has a reflection value known as "albedo." This value defines the "brightness" of reflected light from a world's surface compared to the amout absorbed. A "perfect albedo" would be 1.0, meaning that 100% of the light from the sun is reflected back into space, a value which physically cannot exist.

Earth has an albedo of 0.3, meaning that 30 percent of the sun's light is reflected back into space while the remaining 70% is absorbed into the surface by plants, ground and even water.

This balance of heat absorbed versus heat reflected is the basis for climate and weather in a balanced, sustainable world. If the Earth was completed encased in ice, the albedo of our planet would be quite high – 84%. But the primary ice on our planet is confined to the two polar masses, particularly the south polar cap, and some seasonal snow and ice; any drop in those reflective sources, and the albedo goes down.

Thus, the heat goes up.

Additionally, any increase in aerosol gases in our atmosphere will result in absorption of heat energy from the sun, notable among theses is carbon dioxide which is increasing at an alarming rate. Rebecca Lindsay, (State of the Climate: Atmospheric Carbon Dioxide, NOAA/American Meteorological Society, 2018), revealed that in 2017 the amount of carbon dioxide concentration in our air is now at 405 parts per million (ppm); the highest that the Earth has seen in at least 3 million years. In the past 800,000 years, the concentration has never been this high (see Figure 7, below)

Figure 7 – The sudden increase in Carbon Dioxide
concentration in Earth's atmosphere
Courtesy NOAA

Perhaps even more alarming by this NOAA study is the increase in carbon dioxide in our air just in the past three decades as shown in the NOAA graphic shown as Figure 8, below:

Figure 8 – The steady climb of carbon dioxide in the atmosphere since 1975

Graphic courtesy NOAA

This combination of decrease in reflectivity of the Earth's surface from reduced amount of surface ice and the increase in heat-holding gases such as carbon dioxide results in a very dim outlook for the future of the climate of Earth.

Fast Forward to Climate 2020

Climate change is not myth. Global warming is not science fiction.

The daily average worldwide temperatures are being measured by many sources. According to an ongoing temperature analysis at NASA's Goddard Institute for Space Studies (GISS), the average global temperature on Earth has increased by about 0.8° Celsius (1.4° Fahrenheit) since 1880.

It may not seem like a lot on your thermometer, but this – coupled with the out of control release of greenhouse gases into the atmosphere – has resulted in catastrophic and perhaps irreversible consequences for the future of Earth.

Eighteen of the nineteen warmest years on record worldwide have occurred since 2001, with 2016 being the hottest on record. Each year, the average temperature bumps up ever-so-slightly, just slow enough that most people do not notice. A one-degree *global* change is significant because it takes a vast amount of heat to warm all the oceans, atmosphere, and land by that one degree. In the past, a *one- to two-degree drop* was all it took to plunge the Earth into the **Little Ice Age**. A five-degree drop was enough to bury a large part of North America under a towering mass of ice 20,000 years ago, the Laurentide Ice sheet of the last major ice age.

So let us stop denying all of the climate change evidence – stop pretending that some "Green New Deal" will make any difference whatsoever this late in the game of climate change – and examine what we are faced with today, because the human race in its endless population explosion and greed did not pay attention in the 1970's when efforts to reduce both population growth and greenhouse gas emissions were largely ignored.

What do the coming years look like as we journey ahead in our present path of human consumption and proliferation? Let us look at the recent years of 2018 to 2020 to see what might be in store.

2019 – European Heat Records

In early summer 2019, record heat began to scorch most of Europe and places normally minimally affected from intense heat index readings and suppressing waves of humidity and heat. France, Germany, Poland, the Czech Republic and Spain – along with most all others in the European community began to realize what was to be the hottest summer of record. As early as June nearly every monthly and annual temperature records – record highs – were broken.

Deaths from heat have been recorded in places that never experience heat-related mortality, while heat rose to an incredible 105.6 degrees F in Clermont-Ferrand, France, this only to be topped by that of the small town of Villevieille where it reached a whopping 113.2 degrees on June 27, 2019.

French President Emmanuel Macron mobilized his "entire government" to deal with the unprecedented heat wave, including the closing of some public events and schools; the FIFA Women's World Cup soccer matches went on as planned.

The heat is starting to spark wildfires similar to those so familiar to California in the western United States; the worst fire in two decades has burned over 10,000 acres in Catalonia, Spain.

Compounding Europe's ability to deal with this surging heat is the population density and the fact that nearly all population now resides within urban areas with buildings, roadways, structures....steel and concrete and no relief from green and cooling forests. Climate scientists are stating fairly conclusively that this situation is only getting worse, as seen from the 2003 heat wave and the previous 2018 heat wave.

Estimates of deaths from heat in Europe in 2003 were between 30,000 and 70,000 persons, the hottest period ever recorded since the 16th century. Heat stroke and the simple difficulty of breathing in the heated air affected the elderly far more than any other faction of the population.

Europe, because of the normally mild summer temperatures, is particularly prone to heat related fatalities; Germany for example has only an estimated two percent of its homes air conditioned. The departure from the normal expected heat can be particularly harmful to those not expecting such searing and long-lasting heat.

Figure 9 – Actual air temperatures on June 27-28 2019 in Europe

Scientists in Europe, basing their speculation on the sudden change in climate and temperature on that continent, surmise that similar heat waves should be expected on a frequency five times what used to be considered normal. Even infrastructural efforts, European governments are researching ways to combat heat decay that can affect bridges, towers, rail tracks and other essential structures for human activity. Because of the fear of buckling, new restrictions have been placed on the hitherto untethered driving along German's famous *Autobahn*.

Other Regional Examples

Because of the Earth's axial tilt, when summertime blazes forth in the northern hemisphere, it is winter in the southern continents and lands. Even still, the winters in Australia and other south hemisphere countries and mild and even warm. Summers, by comparison to those of the northern lands, are equally brutal and record heats are piling up years down under.

In the north, Greenland blasted to over 40 degrees warmer than normal for days in June 2019, this accelerating the melting ice sheet that covers most of that land. Extreme heat in India has killed hundreds in 2019 with record temperatures recorded for prolonged periods longer than ever before seen.

Even within the Arctic Circle in 2018 were dozens of wildfires cause by hot and dry conditions not normally experienced in those mostly frigid lands.

Baked Alaska

No where else can compare via contrasts the scorching heat to Alaska; normally a mild and even cool getaway from the summer extremes, the entire extent of this huge state was under a heat wave throughout the summer of 2019.

On July 4, temperatures rose to nearly 100 degrees in many major towns and cities in Alaska, resulting in hundreds of spotty wildfires surging to crews unfamiliar with searing heat. At the Anchorage International Airport, the high hit 90 degrees, far in excess of what meteorologists thought would be a record never to be broken – 85 on June 14, 1969.

The average high temperature for this region of Alaska is only 65 degrees.

The Fourth of July celebration was seen without fireworks due to the extreme conditions of 2019, with the state Fire Marshall banning the sales of fireworks because of extreme fire danger from very dry and hot conditions. Among the 23 fires that scorched the magnificent landscape of Alaska was the Swan Lake Fire, which consumed over 100,000 acres of land.

On a somewhat encouraging note is that at least Fairbanks, Alaska, has experienced all this before; in 1919 the high temperature on July 28 hit a whopping 99 degrees, and again in 1994 when it flirted with the mid-90's. Fort Yukon, located in the central part of the state, recorded a temperature of 100 in 1915 for the state record.

Courtesy Accuweather.com

Going Green or Turning Blue?

All of this is foreboding to say the least. It is depressing....and it should be. Sixty years ago there were a handful of scientists who predicted that if unchanged our world might accelerate climatologically into an irreversible state. Sixty years ago, the world had its hands full of the Vietnam war and other diversions that made "the future" seem like just that: a fantasyland in which other people lived.

But this was not a myth.....this is reality and not science fiction.
We are facing the reality of a world growing warmer by the year and hotter by the century to a point where the human race may not be able to survive the conditions.

Every aspect of this is caused by humans – none of the other billion species of living things on this planet has caused this, only humans. This point is covered thoroughly in Part I of this publication. We have destroyed ecological and climatological balance through two oversights:

1) overpopulation and failure to curb the infinite demand for finite Earth resources;
2) destruction of the environment that provides for stable climate, clean air and water, and unlimited resources.

Following is a brief scenario (such an over-used word today) of what likely will take place following our unwillingness and/or inability to rectify the cataclysm of global climate change.

The Heat Goes On – the presence rate of increase of temperature increase is moving from a linear growth to an exponential climb; this is caused by two factors:

1) Overpopulation – this requires the removal of trees and other oxygen-providing plants to provide room for houses, roads, parking lots, commercial buildings excavations, harvesting of resources, etc., while at the same time contributing to the over-abundance of carbon dioxide which contributes to greenhouse warming.
2) Emission of Greenhouse gases – in addition to carbon dioxide emissions from breathing of billions of humans, we are pouring harmful heat-retaining gases from industrial waste, vehicle emissions and burning at an ever-increasing rate. Carbon dioxide has increased *40 percent*; methane *160 percent*; nitrous oxide *20 percent*. Each of these gases contributes to the retention and

transfer of heat within the Earth's atmosphere; together the form a catastrophic barrier to prevent radiative cooling the Earth into space.

Glaciers Melt – the increase in atmospheric heat retention causes the oceans to warm and the air to warm, both factors in glacial degradation. Greenland has virtually lost all of its glacial ice and will likely never regain that; the northern polar regions' ice thickness has depleted to less than one-half. The potentially potable fresh water locked up in glacial ice is virtually lost once it merges with sea water. Worse even is the reflective effect of the glacial ice is lost, thus causing drastic increase into the already warming blanket of air captured by the Earth's atmosphere.

Evaporative Index Increases – the increasingly warm atmosphere begins to accelerate the evaporation of the lakes, streams and oceans and thus disrupts the nature water cycle. The increased evaporation leads to locally torrential rains, violent storms and regional weather pattern uncertainties. In many locales, increased evaporation cannot be followed by rain because of the upward flow of heated air from the super-heated ground thus the moisture becomes locked into moisture-laden clouds.

Clouds increase – as the moisture-laden clouds rise rapidly upward from heat-induced rapid evaporation of water sources, the clouds accumulate and thicken; one would assume at first that the thicker clouds would result in cooler temperatures, but just the opposite occurs.

Solar Radiation Absorption Increases - the *Greenhouse Effect* demonstrates that sunlight enters the atmosphere as high energy **ultraviolet radiation**, is absorbed by the ground and re-radiated as a much weaker energy – **infrared**. The less energetic wavelengths of infrared are trapped by the Greenhouse gases and the moisture in the clouds and the heat is encapsulated by the Earth's own air.

Temperatures rise per Greenhouse effect – not only does the ground absorb the solar radiation and result in radiative warming, but likewise does the air and the moisture-laden clouds increase in global temperature. At a critical point the upward-rising warm air becomes a more powerful vortex than the gravitational pull on cooling moisture which normally would fall. In short, it becomes to warm to rain.

Fresh Water Decreases – potable water that once filled fresh water sources becomes locked in clouds, diluted in the salty seas, and becomes unable to recycle/perk as per the natural water cycle for Earth.

Figure 11 – The Earth's Normal Water Cycle
Courtesy NASA

All water sources "attempt to find the lowest ground" and gravity will cause water flow to ultimately outsource at sea level. Once any water that falls is allowed to flow into the ocean or other salt water body, it becomes unsuitable for consumption. One major component of global warming and climate change is that of localized or regional storms that have such massive updrafts as to cause serious sudden flooding and torrential rains; such rainfall is typically of little benefit since most runs off and is not absorbed or retained as potable water.

Carbon Dioxide Increases – (along with the other greenhouse gases noted on page 15); particularly troublesome is carbon dioxide because of human population growth and the disruption of the cycle of photosynthesis. As described, human population requires housing, roads, buildings and large open areas which are cleared of green plants, the very source of most present day oxygen and one major factor that once kept the increase of carbon dioxide in balance. The green plants utilize carbon dioxide as their resource to make oxygen. Two deciduous trees can provide a year's worth of oxygen for one person, and they consume an even great amount of carbon dioxide in the process. But we are destroying forests and jungles vital for the photosynthesis cycle and for the cleansing of greenhouse gases from the environment.

Plants Decrease from Ecological Balance

Gases percentages (O and CO2) begin to fail – the photosynthesis cycle begins to be insufficient to provide oxygen for an increasing human population, and the carbon dioxide from breathing as well as from industrial activity increases as human activity continues to indiscriminately remove green plants from the landscape of the planet. All the while, the cooling effect of plant life diminishes, contributing to the rapidly increasing level of heat in the surrounding air.

Human Population Continues Unchecked – as the environmental balance rapidly declines, the atmosphere warms and the lack of potable water increases due to unpredictable rainfall and careless human waste. All the while the number of humans exponentially increases on the surface of Earth; so many people at this time – nearly 8 billion – that even with no-growth population (every couple would conceive only two children), the number of humans on the planet within the next century will smother out the available oxygen and water for all to survive.

A Grim Conclusion

There is absolutely nothing "pretty" about the picture that I have just painted. In fact it is disturbing....it is quite ugly. But it is the truth about the state of our planet and the future of the human race.

I have written this because fifty years ago I began traveling, lecturing, and explaining the upcoming downfall of the human race if climate change and over-population was not examined and some efforts were made to curtail both. At that time, these things seemed like that myth – like that science fiction that I keep referring to – to all who heard my lectures. Some people politely listened, nodding their heads as if to agree, others suggested that all this climate change and human population talk was nothing but poppycock and went about their ways of life.

At that time such topics as this were in the future – they WERE science fiction in 1965. People either assumed that there was nothing they could do, or that none of this mattered because frankly, they would all be dead by the time that any of this could come to fruition.

So – today – we have reality. The reality of a super-heated planet that perhaps can become incapable of supporting human life is approaching rapidly and within a few years may be upon us. The reality that we have simply "had too many children" has not sunk in to the masses of humans yet. In fact, celebrities are making a point that families should once again have five to eight children to have "complete lives."

Children are wonderful to spawn in this sea of life, but we are doing so without giving any thought to what this life may be for them in only a few years. For one, I do not want my children, nor their children, gasping for every breath, competing for ground on which to live, and sweltering to die in the heat of an arid world in which a soothing rainfall at night is nothing more than science fiction.

Other Publications by P. Clay Sherrod

Available by contacting ASO at www.arksky or direct through drclay@tcworks.net

Also available through *Amazon.com* or *Barnes & Noble*

Sherrod, P. Clay, *The Arkansas Sky Observatories' Supernova Search Atlas and Guide*. A study guide for the discovery of extragalactic supernovae. 2018, 164pp. Large format soft bound. 350 galaxy photographic atlas. $19.95

Sherrod, P. Clay, *The Genesis Project: Creation and Science through Art* / Premium Collectors Large Format Edition / Hardbound coffetable edition full color – LuLu Publishing. Original artwork and philosophical discussions of Biblical Genesis by author   Arkansas Sky Observatories. $95

Sherrod, P. Clay, *The Genesis Project: Creation and Science through Art* – Arkansas Sky Observatories – Premium hardback edition, full color - Original artwork and philosophical discussions of Biblical Genesis by author. 68 full color pages.   Arkansas Sky Observatories. $34.95 (hard)

Sherrod, P. Clay, *A Complete Manual of Amateur Astronomy,* Simon and Schuster, New York. 1982. Hardbound, softbound.

Sherrod, P. Clay, *The Constellations: Tours for Computerized Telescopes Vol. I* 2017, Lulu Press, Study Guide packed with illustrations and photos; 396pp - Premium paperback - $19.00

Sherrod, P. Clay, *The Constellations: Tours for Computerized Telescopes Vol. II* 2017, Lulu Press, Study Guide packed with illustrations and photos; 550pp - Premium paperback $29.00

Sherrod, P. Clay, Telescope Rx. 2017, Lulu Press, 1st ed, 352pp Large format study and self-help guide; Premium paperback ed., $29.50

Sherrod, P. Clay , *Catalog of Cometary Orbits, Arkansas Sky Observatories* - 2017 , large format, Premium Paperback edition - 296pp. $29.95

Sherrod, P. Clay, *Footprints of Fallen Giants – Pathways to Extinction* / (Dinosaur evolution philosophy and discussion) large format publication of original manuscript with notations, Premium Paperback edition with many original illustrations throughout – 314pp 2016, Lulu Press - $29.95

Sherrod, P. Clay and John Berky – *In Vitro Toxicity Testing: For Carcinogenesis, Mutagenesis and Toxicity* – Franklin Institute Press for the US Food and Drug Admin. 1977, 433pp., paperbound. Available on auction sites and through Amazon. $120+, current market price

Sherrod, P. Clay, *Motifs of Ancient Man*, Rock art and pictographs of Arkansas. University of Arkansas Press, 1983. Some copies available via Amazon and rare book sellers. Paperbound, 160pp. Illustrated. $125+ current market price

Sherrod, P. Clay, and M. Rollingson, *Surveyors of the Prehistoric Mississippi Valley* – *The Toltec Module.* The incredible engineering of the ancient mound builders revealed. 1987, Arkansas Archeological Survey Research Series #28. 162pp. Paperbound – Available through U of A, Amazon, Toltec Mounds and Cahokia Mounds State Parks

Sherrod, P. Clay, *Rocks of Prehistory: The Gumlog Creek Rockshelter* – Premium paperback edition, many illustrations and discussions of prehistoric occupation and anthropological overview of regional Native Americans. 112pp. Lulu Press, 2016 $18.00

Sherrod, P. Clay, *Ancient Quests of the Double-Peaked Mountain: The Gunbarrel Petroglyph* – A captivating discussion highlighted with fascinating illustrations throughout. 72pp. Premium paperback edition. Lulu Press - $15

Sherrod, P. Clay, *You Twit Face: Your Complete Guide to Nomophobia* (our obsessions with the "Smart Phone") – Premium Paperback edition - Hardback edition, 234pp $29,95; soft $19.50 Lulu Press, 2016

Sherrod, P. Clay, *Life Lessons from the Naked Boy.* 2016, Lulu Press, Hardcover edition , Premium Paperback edition - 266pp - $19.95 soft; $29.95 hard cover

Sherrod, P. Clay, *The World's Last Peanut – a Curious collection of politically incorrect opinions and thoughts.* 2018, Arkansas Sky Observatories Publications. Softound, 130pp. Lulu Press.

Sherrod, P. Clay, *Cookin' In The Woods.* 2018, Arkansas Sky Observatories Publications. Spiral Bound cookbook, 86pp. soft cover, $14.

Sherrod, P. Clay, *Fireflies From Orion: The Poetry of P. Clay Sherrod* . 2019, Arkansas Sky Observatories, softbound, 349 pp.

Sherrod, P. Clay, *Fission Population* . 2019, Arkansas Sky Observatories, softbound, 37 pp.

AND MORE.....please see the Arkansas Sky Observatories' Publication Tab at: http://arksky.org/publications

www.ingramcontent.com/pod-product-compliance
Lightning Source LLC
Chambersburg PA
CBHW021016180526
45163CB00005B/1974